How To Be
Energy Efficient

How To Be Energy Efficient

H A Rasheed

iUniverse, Inc.
New York Lincoln Shanghai

How To Be Energy Efficient

iUniverse books may be ordered through booksellers or by contacting:

iUniverse
2021 Pine Lake Road, Suite 100
Lincoln, NE 68512
www.iuniverse.com
1-800-Authors (1-800-288-4677)

ISBN-13: 978-0-595-39263-6 (pbk)
ISBN-13: 978-0-595-83659-8 (ebk)
ISBN-10: 0-595-39263-6 (pbk)
ISBN-10: 0-595-83659-3 (ebk)

Printed in the United States of America

To Dennis Cain Fleck and all his brothers and sisters

Contents

Part One

Introduction

Chapter One

Definitions

The Problem with Money

I can exchange a shirt for a pair of pants with you and we might call it an even trade. To be consistent with our units of measure, we convert the value of both attire to money: The shirt probably cost $20 to $30 and the pants about the same at the local store, and so we can say the transaction was like exchanging $25.

The convenience of money has made it a standard unit of measure into which the value all kinds of products and services are converted, or in a more precise term "equated." Unfortunately, the process of equating everything to money is relativistic. For example, the Indian who sold Manhattan to the pilgrim from England for $15 in beads probably valued them as priceless and the pilgrim valued Manhattan in more or less similar terms.

This makes currency a poor way to measure efficiency.

Example: If you ask an owner of a 2005 Chevy Blazer how far he can travel using $2.50 in gas, he will probably say 31 miles. An Egyptian who owns the same SUV and the same amount of money would answer about 189 miles. Since we are testing the same vehicle, it is difficult to determine how a six-fold increase in efficiency materialized.

Thus it is hard to pinpoint the rock bottom value of anything using currency. Trying to measure efficiency using money is like comparing apples and oranges (as my seventh grade math teacher insisted I couldn't do.)

Definitions

I chose the calorie as a unit to measure energy because if is a common term used on almost all products at the grocery store. But you may ask: What is a calorie, exactly?

Definition 1:
A calorie is a unit of heat required to raise the temperature of one gram of water by one degree Celsius from a standard initial temperature

The other term used in this book is efficiency. Efficiency can be defined in many different ways but I will use the relationship between what we put into a process and what we get out of it. For example, if I put a rod of steel that has been heated up with one hundred calories into a glass holding one hundred grams of water my expectation according to the definition above is that the water temperature would rise one degree from 50 degrees Celsius to 52 degrees Celsius. To express this transformation, I divide the result or output of the experiment by the input which is 100 calories (50 grams of water times two degrees Celsius each) by 100 calories (released by the steel rod) giving and efficiency of 1.

I should add here that I often express efficiency by the term "percent." Percentages are a common way of expressing efficiency and all we do is multiply the efficiency figure by 100. So for the above experiment an efficiency of 1 is expressed as 100%.

Definition 2:
Efficiency is measured as output divided by input multiplied by 100.

Example: If your chicken lays ten eggs but only two hatch, its efficiency is 20%.

Read more in:

Appendix One—The Numerical Perspective
Appendix Fifteen—Conversion Table

Chapter Two

How Much and Where Does it All Go?

Fuel Consumption Category	Calories Consumed
Auto Fuel	1,499,243,965,714
Home Heating Cooling	518,231,419,200
Appliances	134,342,220,298
Interest on home loan	1,309,757,619
Home—All Other	727,643,122
Auto—All Other	311,847,052
Garbage	52,390,305
Water	17,463,435

"Average" Proportional Family Energy Consumption in the Northwest

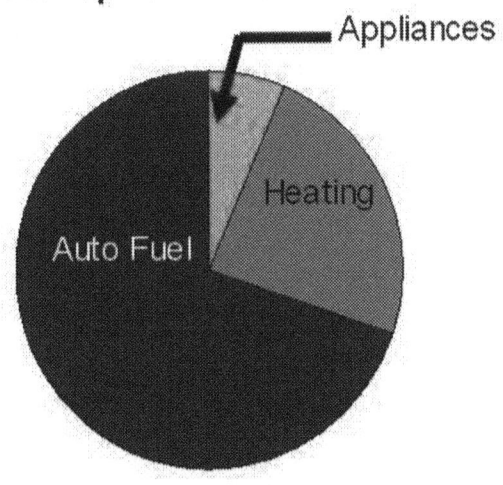

Two people starting on their journey together through life in the Willamette Valley of the Pacific North West have the prospect of buying a 1400 ft² home with a two car garage, each commuting ten to twenty miles to work each weekday. The weather in the northwest is temperate; temperatures in the 80's in the summer and around 35 degrees (Fahrenheit) in the winter. They might see an inch or two of snow for a week or two. The previous tables show the average energy consumed over a thirty year period.

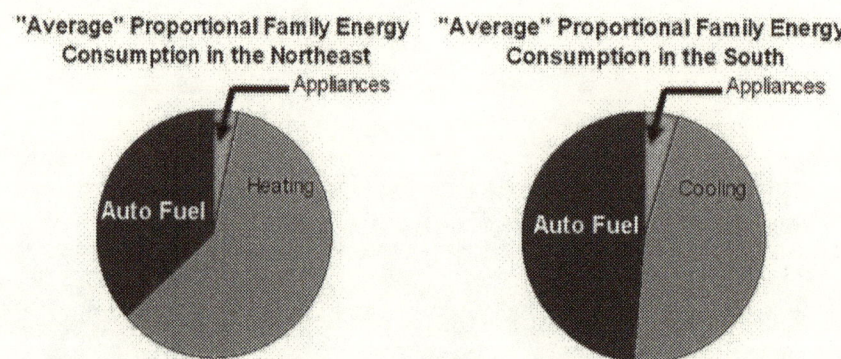

The previous graphs illustrate the proportion of energy used in the three top most energy consuming categories in the northwest, northeast and the south. The next eight chapters will explain how to reduce energy consumption in each those major category.

Read more in:

Appendix Fifteen—Conversion Table

Part Two

Auto Fuel

Chapter Three

Auto Fuel Consumption

Facts

The US imported **19,700,000** barrels of crude oil a day in 2002
This is equivalent to **27,788,406,300,000,000** calories a day

Where does it all go?

Lets start simply. First, consider what it takes to travel a mile. The average person expends about 300 calories walking that mile.

A single mile on a bicycle will required a little more than 300 calories. Although the bike weighs about 40 pounds the difference in performance is negligible because a bicycle conserves momentum: once it starts rolling, it keeps doing so.

The internal combustion engine is a whole different story. Its mere existence consumes about one million calories on a journey of one mile. It is only 20% efficient, and the million calories are consumed by the pistons moving back and forth at high speed, pushing gasses in and out.

Another 230,000 calories are consumed just overcoming the weight and friction caused by the bulk of the vehicle that the motor is trying to move. In addition, at high speeds, the displacement and resistance of the air creates greater and greater drag.

Once the inefficiency of the engine, friction and drag are accounted for, it takes 22,000 calories to move the whole 4–5000 pound vehicle that one mile.

In general physics there are three factors that determine fuel consumption of a vehicle. They are mass, distance and time.

By far the heaviest influence on fuel consumption in our fast-moving, mobile society is distance and time. The longer the distance, and the shorter the time in which you want to get there, the more fuel is consumed.

What this means is that you will spend five times more energy going two miles in the time it took you to go one mile. On the other hand, if you want to travel a mile in half the time of two miles, it will take you about five times as much energy. This exponential increase in fuel consumption is derived from the equation Energy = ½ Mass * Speed^2 where anything related to speed is exponential.

This relationship between time and distance caused some auto makers to place an RPM (Revolutions per Minute) gage in their autos. This gage times the number of revolutions per minute of the crank shaft which has a direct relationship to the turning of the wheels and thus the auto's speed.

In actuality, the efficiency of an auto with an internal combustion engine is much less than 20% because the process of gaining the speeds (miles per hour) that people feel they need, requires revving up the engine (increasing its RPM) and resulting in greater fuel consumption.

In summary, the costs of going that one mile requires the engine consume 80% of its energy before the vehicle even moves. In other words, it is an energy hog.

The fact that internal combustion engines are only about 20% efficient has prompted many to take interest in alternative sources of energy. The electric motor has been around as long as the internal combustion engine and it is about 45% efficient and there have been several attempts at powering autos with it. Unfortunately the laws of physics tend to spoil its chances of success, because the electricity generated to power an electric auto needed to be generated by an internal combustion engine, defeating the purpose.

Energy Usage of an Electric Auto

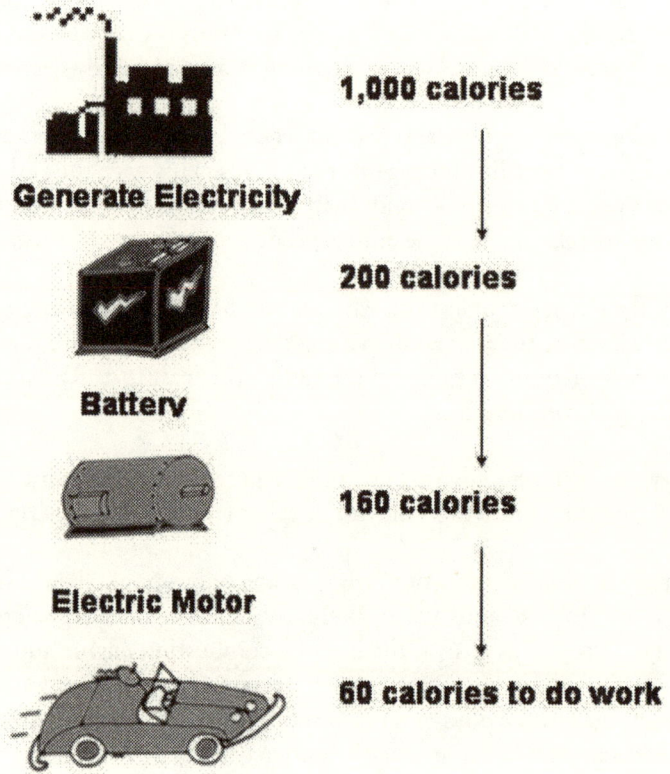

Generate Electricity   1,000 calories

Battery   200 calories

Electric Motor   160 calories

60 calories to do work

The same laws apply to the fuel cell technologies being developed at this time. A fuel cell generates electricity from stored hydrogen and oxygen, and has the potential of taking the place of the battery in an electric car. It has higher efficiency: car batteries are about 75–80% efficient while fuel cells are theoretically 83% efficient.) But again, the stored hydrogen for fuel cells must be produced using an electric current which is only available from electric generators that run using internal combustion engines. The star quality that fuel cells have today comes from the idea that it uses water as a source for its fuel and its exhaust is water which, to the non-expert, looks like that is all there is to the fuel cell concluding that it solves the fuel and air pollution problems at the same time.

Energy Usage of an Auto Using a Fuel Cell

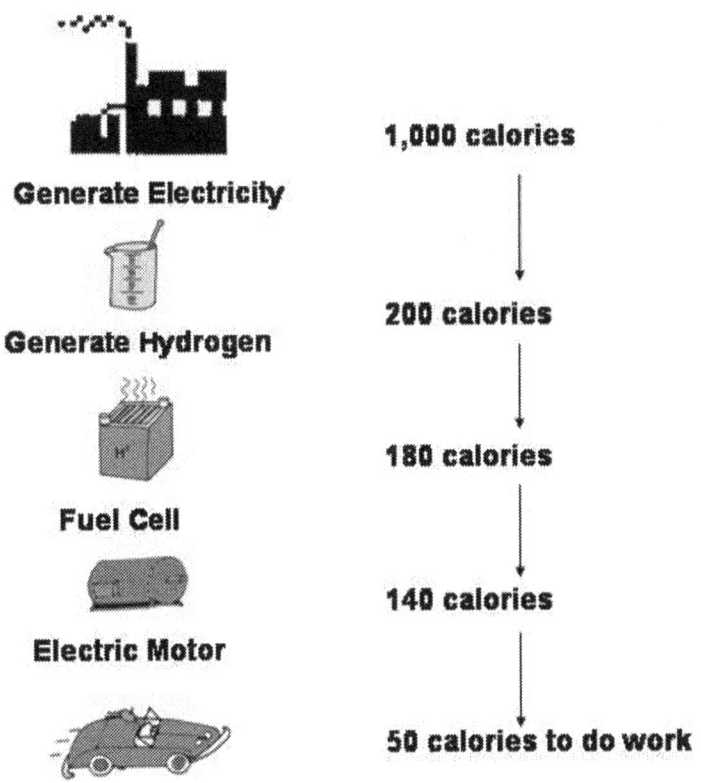

Generate Electricity   1,000 calories

Generate Hydrogen   200 calories

Fuel Cell   180 calories

Electric Motor   140 calories

          50 calories to do work

Of all the innovations that have been attempted, the hybrid auto has a better chance of success. A hybrid auto is essentially an electric auto. The only difference between it and the commonly known electric auto is that its fuel is gasoline. The advantage of a hybrid is that the internal combustion engine runs constantly at optimum efficiency. In stop and go driving, a regular auto revs its engine to high RPMs many times for each mile traveled, which is not the most efficient use of the engine. By reducing the inefficiencies of the internal combustion engine, coupling it with an electric storage system and a more efficient electric motor, the hybrid is able to increase efficiency by ten to twenty percentage points.

Diagram showing the energy stream in a hybrid auto

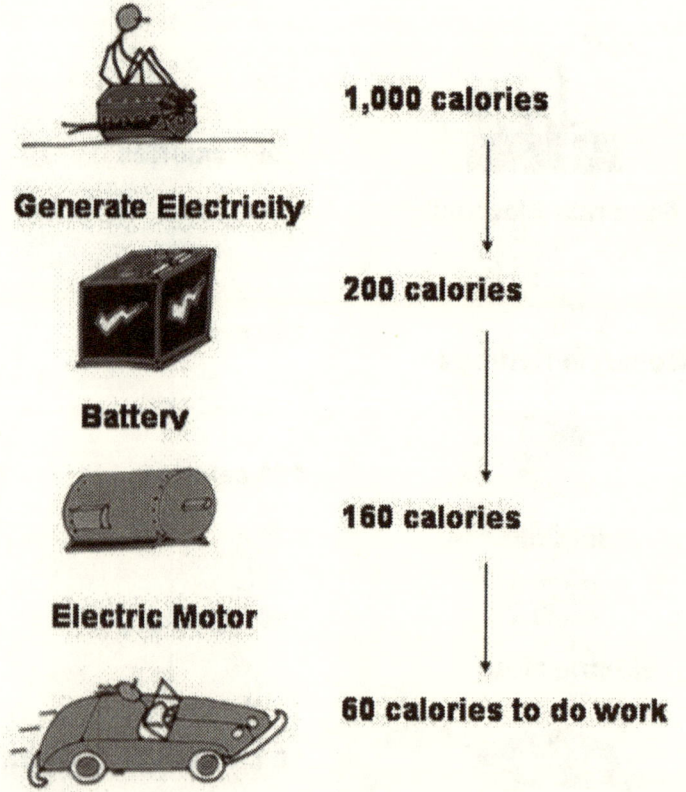

Generate Electricity

1,000 calories

200 calories

Battery

160 calories

Electric Motor

60 calories to do work

Table showing efficiencies of various auto components	
Auto Component	**Efficiency**
Internal Combustion Engine	20%
Electric Motor	50%
Battery	80%
Fuel Cell	83%

From the above table of various auto component efficiencies, it is clear that the internal combustion engine is the least efficient and we are left with the electric motor as the most efficient converter of energy to mechanical performance. Although the fuel cell is more efficient at storing and delivering electrical energy to the electric motor, it costs energy to make the hydrogen required to make it work, which further reduces its efficiency compared to the regular battery.

If we give a vehicle with a battery and an electrical motor 1,000 calories we will be able to provide around 400 calories to do work (as apposed to the electric car that provides 50 to 60 calories.)

But how can we get around using the internal combustion engine to generate electricity? There are several other sources of electricity that haven't been introduced viably on a large scale into the energy market. These are solar-and wind-generated electric energy.

Recommendations to increase your fuel efficiency:

1- If the distance you need to go is a mile or two, walk it

2- If the distance you need to go is two or three miles then ride a bicycle;

3- If you are single, look into buying a motorcycle;

4- When shopping for a vehicle:

    a. Choose a vehicle that fits you like a glove. This means avoid buying one to accommodate casual uses. For example, if you need a four seat car and once in a while it would be nice to accommodate six people then buy a four seat car and rent a limousine on special occasions. The same goes for a four-five seat pickup; you can always rent a pickup when the time comes;

    b. Look at the EPA mileage sticker very closely and select the one with the most miles per gallon from your choices;

    c. If EPA mileage stickers are the same between two choices, chose the lighter car;

    d. Given EPA mileage stickers that are the same between two choices, chose the one with the smaller engine by looking at the engine displacement;

    e. Given If EPA mileage stickers that are the same select the car with the smaller and aerodynamic body;

    f. Electric or hybrid cars are more fuel efficient.

5- If you have a car that doesn't have an RPM gage, install one and try to avoid going over 3,000 RPM while driving;

6- Make as few trips per day as possible by grouping as many of them as you are able to in one outing by taking out a few minutes for planning;

7- Use routes with the fewest stops;

8- Choose the shortest routes;

9- Keep track of the miles you travel;

10- Keep you're tires properly inflated;

11- Keep you're vehicle in the best working order;

12- Invest in solar and wind power plants.

Energy Usage of a Solar-Electric Auto

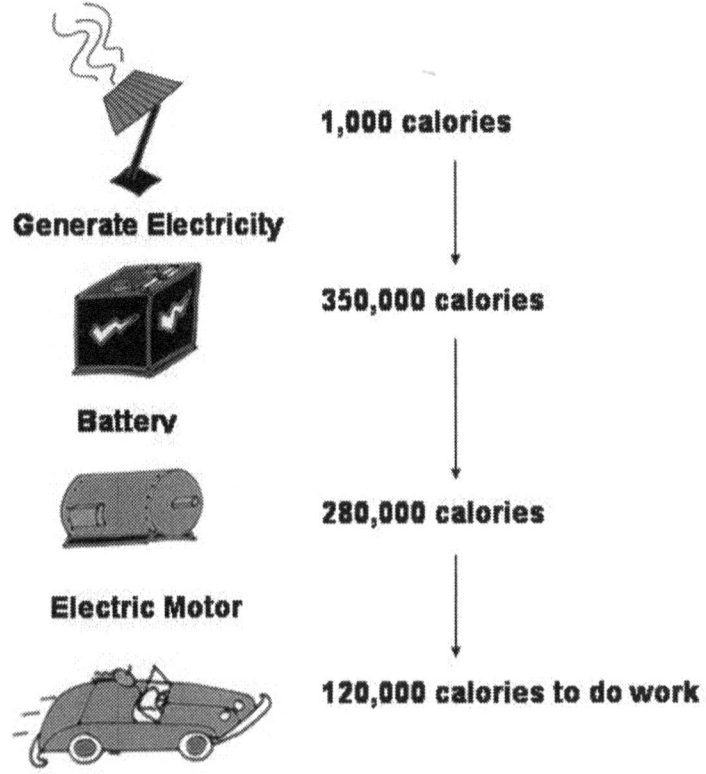

Generate Electricity

1,000 calories

Battery

350,000 calories

Electric Motor

280,000 calories

120,000 calories to do work

The efficiency results may not make sense because the output is greater than the input resulting in efficiency greater than 100%. In this case solar energy is almost free resulting in an efficiency of 350,000%. This is because I am an "open" system and a closed one at this point.

Read more in:

Appendix Four—Battery Considerations
Appendix Five—Fuel Cell Considerations
Appendix Eleven—EPA Fuel Economy Auto Testing
Appendix Fifteen—Conversion Table

Part Three

Heating and Cooling

Chapter Four

Underground Dwellings

The human body is an adaptive machine. It is able to tolerate and change to accommodate many environmental conditions, including temperature. The ideal temperature for the human body is around 70 to 80 degrees. This allows it to function and get rid of excess heat generated through physical and mental processes. The temperatures that the human body can adjust to range from about 50 to 90 degrees. At 50 degrees, moderate physical activity or an additional layer of clothing can restore considerable comfort. At 90 degrees, the body's natural cooling system, sweat, will also increase its ability to rid itself of excess heat.

One option is to relocate if the weather in your current place of residence is too hot or cold. However, this may not be an option.

If you can't move, then consider moving underground. It is well

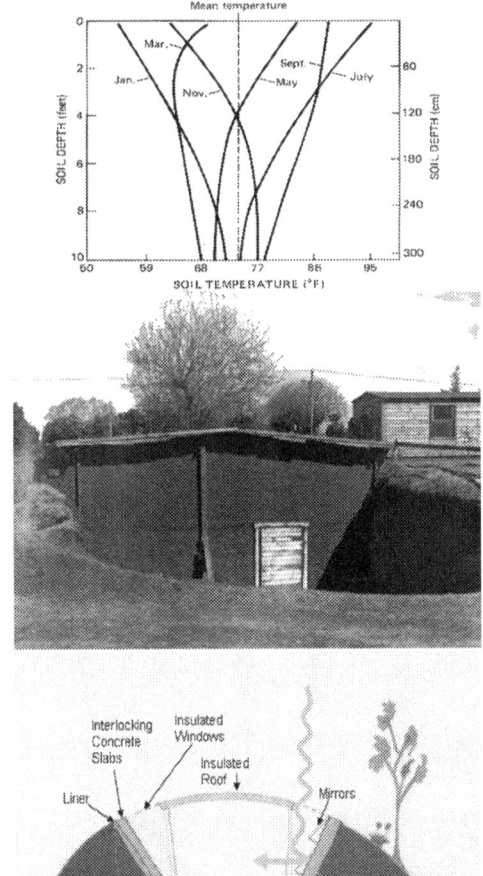

known that the deeper you go the more stable and comfortable the temperature is. For example, in West Texas a study was made of soil temperatures at various depths and the results can be found in the diagram shown at the start of this chapter.

Although the temperature varied from 55 to 95 degrees at the surface, the temperature ranged from a most comfortable 68 to 77 degrees at ten feet down. This eliminates the need for heating in the winter and cooling in summer. This idea is not new and is practiced throughout the world and Native Americans. (See to the fist photo in this chapter, depicting an underground dwelling in Arkansas.)

The third image on the first page of this chapter is of a conceptual drawing of a structure that does not require heating or cooling in summer or winter where the outside temperature ranges from 50 to 95 degrees. This underground structure is built in a mound of dirt to avoid high ground water levels or flash flooding. It is constructed in much the same fashion as an artificial botanical pond.

First, a rubber liner is placed in the dugout area and lined with concrete slabs at the angle of repose. The living area is then built within this cavity with a well insulated roof. The gap between the conventional square structure inside and the slanted walls is covered with insulated rain-proof windows. These windows allow much-needed light, and in this particular structure, the light from the sky is reflected into the living cavity by the use of mirrors. Many variations on this basic theme are possible.

The Shape of Things

One observation from nature is that reducing the surface area exposed to the outside elements conserves the effort of keeping its interior cool or warm. This ideal conceptual form is the sphere.

Some innovations have come from this, one of which is the geodesic dome structure. To the right is a conceptual drawing of a geodesic dome living structure consisting of two floors.

These structures became very popular in the 70s and 80s and were cost effective to make and build. Some come in kits that one or two people can put together. Besides conserving energy, they are strong and durable, withstanding storms that other structures can't. Today, kits can be acquired made of foam for simple installation and supper insulation.

But say that you aren't comfortable with an underground or dome-shaped house use some of the ideas these structure bring to the energy conservation arena, like building a house as close to a cube shape as possible.

The following are conceptual drawings of one, two and three level structures all with 5,000 square feet of living space. Please note the reduction in surface area from one, two and three levels.

One-story structure with surface area of 13,000 sq. ft.

Two-story structure with surface area of 9,000 sq. ft.

Three-story structure with surface area of 6,800 sq. ft.

As can be seen from the above drawings, energy efficiency can be improved by 45%. For example, if energy costs are around $200 a month during the winter, moving from a 5,000 sq. ft. one-story house to one that is the same square footage but is three levels could reduce energy costs to about $110.

The diagram to the right by Leonardo Davinci illustrates that we don't really take up square spaces, but round ones.

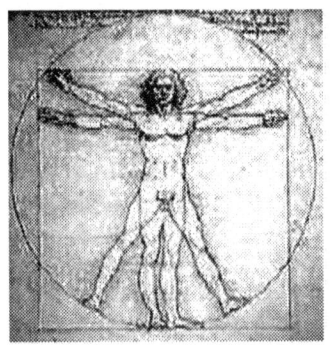

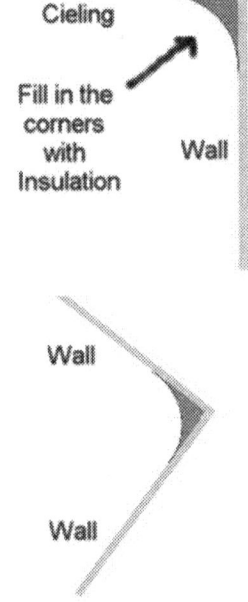

For example, the corners between the ceiling and walls in our living quarters are essentially useless and increase the surface areas of the house to the outside elements. We could potentially reduce energy costs by 5% or more by installing a foot-by-foot right triangle of foam in the these areas.

It isn't practical to treat the corners between the floor and the wall in a similar manner because our furniture is basically square in shape.

There are other benefits in building houses in the shape of a cube in that decreased surface area translates into less insulation required, as will be demonstrated in coming chapters.

Chapter Six

The Sun

The sun is an important source of energy in the form of heat. In the north it is welcome anytime, but in the south it can be a burden when it comes to home comfort. The following discussion will start with over-exposure to the sun and what we can do about it. At the end of the chapter, we will talk about the welcome sun in the north and how we can capture its benefits.

The problem with the sun in the south is that it is too intense especially when it hits our homes directly. Most houses have five sides above ground, and luckily the roof works like a buffer to heat if the house is designed properly.

Originally the roof was designed to protect the structure from rain. In doing so it creates a cavity over the house. These cavities are supposed to be well-ventilated to prevent problems of moisture buildup and resultant complications such as mold and rot. Unfortunately, the degree of ventilation for moisture prevention doesn't solve problems with heat. It is common for the attic to reach temperatures 150 degrees which can make living under it unbearable. A simple solution to this problem is to increase the number of roof vents to let the heat

escape. Along with the heat escape vents, you must have sufficient air inlet vents at the base of the roof to allow cooler air to replace the heat.

There is an array of vents available ranging from $10 to $150. The more expensive ones have fans that can be turned on and off depending on the temperature and or humidity of the attic. Some even have fans that operate on electricity from solar panels regulated by thermostats.

Examples of roof vents and fans

Reducing the direct exposure of the house to the sun can result in a 50% to 90% decrease in the amount of energy (heat) that it is exposed to, depending on the surrounding terrain.

One way to reduce exposure to the sun is by building homes with the least exposure to them, with the narrow end of the house facing the sun, as in the following diagram.

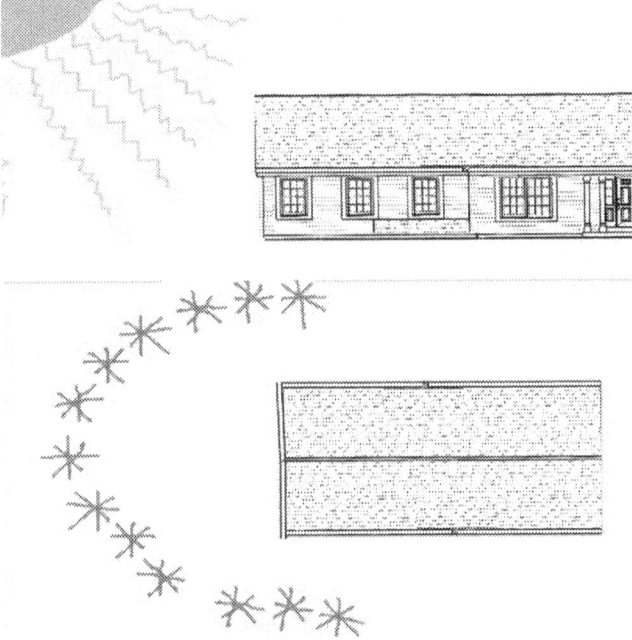

To further decrease the sun's power we can plant trees or shrubs between the house and the sun, so that only indirect sunlight reaches the house, as can be seen in the above drawing.

Evergreen trees are a good choice if the winter months are also hot. To pick a suitable tree for such a project, you need to know the angle of the sun from the horizon; this will dictate how tall a tree to plant and how far from the house it should be planted.

To the right are drawings illustrating how to estimate the height of a tree from your house to provide maximum protection from the sun's direct rays for one, two and three story houses.

Please note that the protractor is placed at the roof line and that the tree is placed well away from the structure. The farther out the tree the taller the tree should be. Caution should be taken when placing a tree close to a structure because of damage it may cause due to mold, etc. Always consult your local government.

The reason why you should estimate the height of the tree from the roof of the house is to make sure that the whole side is shaded. Otherwise, the top floors may not get the full benefit.

Another method of reducing the exposure to the sun is to building a porch, arbor or trellis near the house. They can support vine plants like ivy and grape, as illustrated in the first drawing to the right.

You can create indirect sunlight with shades on the sunward side of the house, as in the second drawing or the roof line to create shade.

When planning the use of vegetation to create shade, you must also take into consideration the effects it might have on air circulation and make sure that there is enough air movement to prevent the establishment of mold and mildew.

As we move north winters are colder, consider planting deciduous plants (plants that lose their leaves in the fall) to protect the house from the hot sun in the summer and increase exposure to it in the winter, to help with heating bills. Instead of pine trees, plant maple, and instead of ivy, plant grapes.

In addition, shade structures can be built to protect the house from the sun in the summer and expose it to the sun in winter. A porch or other shading device can cover the exterior walls of the house in the summer only. As the sun is lower on the horizon in the winter months, rays can sneak under the shade structure.

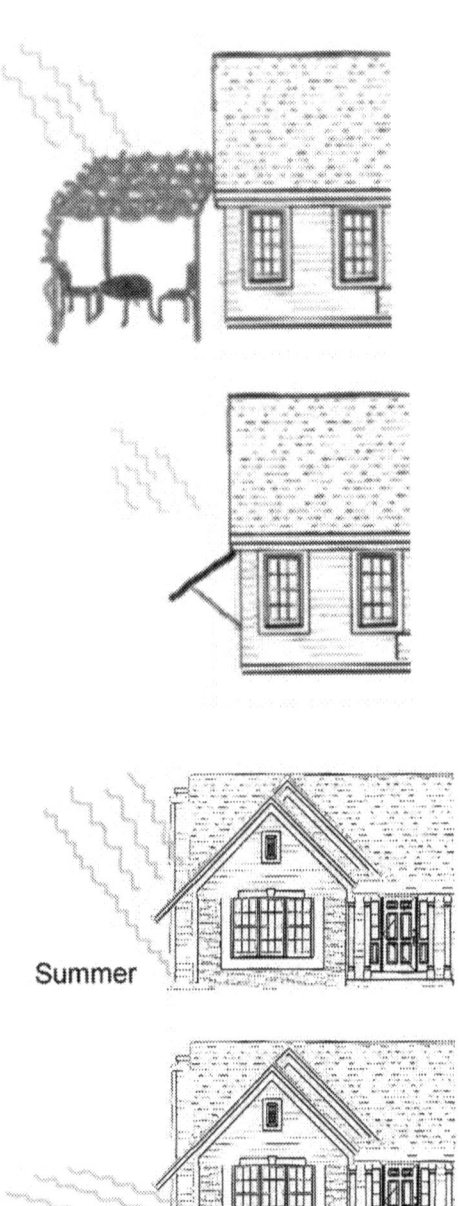

Summer

Winter

Far north, it may be that you do nothing and allow the full impact of whatever sun you have to reach your house.

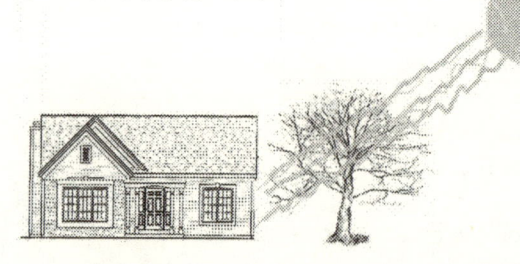

Increase breeze velocity using trees you planted for shade with few lower branches, or trimming the low growing branches yourself.

In planning the position of a tree on your property, it is also beneficial to consider its effect on wind and breeze. In hot climates, it would be advantageous to have trees that allow the cooling effect of breezes. For single-story houses in hot climates, consider planting trees with few low branches.

In cool or cold climates breeze and wind may increase your cost of heating due to the wind chill factor. In this case, consider planting trees with low-lying branches like evergreens or shrubs that break the effects of the wind.

Read more in:

Appendix Eight—Planting Distances for Trees to Shade a House (in feet)

Chapter Seven

Paint Colors

Walls exposed to the sun can be treated to reduce or increase the absorbtion of the sun's energy. In the hot south, the desire to prevent heat from building up in our houses is opposite to those in the cold north, where the sun's rays are a welcome sight.

Generally, light colors reflect heat, while darker colors absorb heat. So it may be wise to paint your sun-facing outside walls white, light blue or light green to keep the walls cool and prevent the migration of heat from the walls to the interior of the house.

In chillier climates choose dark colors to squeeze every calorie out of the precious rays that appear.

The "do-it-your-self-ers" building a house can do some experimentation with color by letting a box sit outside for an hour or two with a thermometer inside. Note how hot it gets with different paint colors you intend to use. You'll be surprised at the effect of paint color have on the temperature inside the box.

A better experiment would use a box constructed of the actual materials used in your final product (drywall, insulation, siding, etc.) To extrapolate your results to the finished product, use the following formulas in the order they appear, and carry the coefficient forward for each measurement:

Coefficient (Cx) = (time exposed **times**

width **times**

height **times**

temperature change **divide by**

volume) of box

Change of temperature in house = Cx **times**

(volume **divide by**

(exposure time **times**

width **times**

height)) of house

(Where volume is width **times** height of walls **times** depth

If you have a dark-colored house, an air conditioner, and you decide to paint your house a lighter color (say, yellow), you can expect to have a return in calorie investment of around 1500% every month in the summer! (Assuming it took you 4 gallons of paint to cover 50 feet of exterior sun-exposed walls of a 1500 ft^2 R17 insulated house, and that it reduced your interior house temperature by about three degrees.)

Read more in:

Appendix Nine—Light
Appendix Ten—Paint Color

Chapter Eight

Windows

So far we have talked about the sun's rays being converted to heat. We also talked about heat migrating into the interior of the house through the walls. Windows are transparent and allow sunlight directly into the house. This can have good or bad consequences, depending on where you live.

A load bearing wall may have no windows, other walls can have 50% of the surface area in windows. In hot climates we need to prevent this energy from entering the house. The simplest way to do this is to place curtains on the outside of the window. It is a misconception that we need to place the curtains on the inside: energy becomes trapped in the house. Outside curtains trap the energy outside and it simply is diffused into the outside air.

How curtains convert the sun's energy to heat and what happens if the curtains are on the inside or the outside:

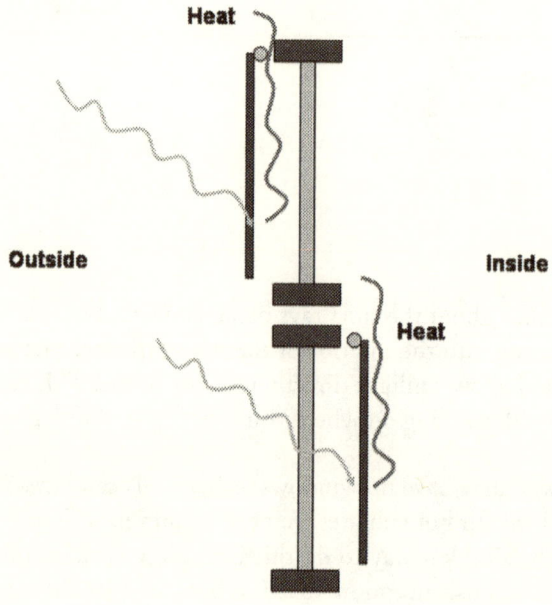

In colder climates, placing dark curtains inside will trap the sun's energy in the form of heat inside the house. You don't have to limit your efforts to trap the sun's energy to curtains. Painting the interior of your house in dark colors will have a similar effect, and allow you to enjoy the sunlight as well.

You can purchase insulated windows that can prevent the migration of heat from the outside and vise versa. Despite the short life expectancy of these windows (about five to ten years) they are well worth their high cost.

A comparison between aluminum and well insulated widows for a 1,500 ft² house located in the temperate northwest demonstrates that in the first year, a conservation of about 3 billion calories is possible, resulting in an efficiency rating of 4000%.

Read more in:

Appendix Nine—Light
Appendix Ten—Paint Color

Chapter Nine

Insulation

When the temperature outside is 20 degrees and all your doors and windows are closed, chances are, your house will become the same temperature through conduction of heat from the interior to exterior.

The best way to handle this problem is by the use of insulation, which hinders the conduction of heat. This conserves heat so you run the furnace or air conditioner less often.

Insulation comes in several forms, but we will talk about fiberglass here simply because it is one of the most popular materials. It is rated by "R values" which. reflect its ability to resist the conduction of energy, the higher the "R" rating, the greater the insulation's ability to resist the conduction of heat.

Below is a table illustrating the increase in efficiency as the result of insulation where calculations are based on a delta temperature (the difference in temperature between inside and outside the house) that lasts 10 hours a day for 30 days, and assuming that windows cover 30% of outside wall area of a 1,500 ft² house. As you can see, you can't go wrong.

Efficiency after Installing Insulation			
Delta Temperature °F	R13	R19	R25
10	598%	265%	73%
20	1,195,553%	529,039%	146,541%
30	27,029,931%	11,960,899%	3,313,097%

The drawback to insulation is that it seals the inside quite tightly and may cause ventilation problems. However, there are ways to ventilate the inside of the house without opening doors or windows and defeating the purpose of insulation. A popular way to ventilate and conserve heat or cold is to use a heat exchanger. Heat exchangers have been use for a long time in industry but haven't been widespread in homes because of their higher cost in comparison to the price of oil.

High performance heat exchangers can deliver 90% or more heat or cold conservation (or a 10% or less reduction in insulation efficiency.)

Read more in:

Appendix Six—R-Value Considerations
Appendix Twelve—Heat Exchangers
Appendix Fifteen—Conversion Table

Chapter Ten

Heating Water

In the south, you can almost do without a water heater by using a solar water heater and save your usual utility energy costs.

A solar water heater consists of a solar collector and a water tank. The collector is made of an insulated window box with coils of copper pipe on a copper sheet supported by a strong flat support which are painted black to maximize the absorbtion of heat from sunlight.

As the water in the copper pipes heats, it moves to the insulated water tank, providing a source of hot water to the house night and day.

As you move north, the angle of the sun and the presence of clouds reduce the amount of heat that can be absorbed and thus the availability of hot water, especially in the winter. A solution to this problem is to couple a solar hot water system with an ordinary water heater using a heat exchanger so that the later can warm the solar heated water up your desired temperature, while still saving you significant water heating costs.

There are many designs to solar water heaters, from simple plastic units that hang on a roof, to more elaborate ones consisting of copper or brass tanks hidden in the roof, with the solar collector fastened on top.

The key to any solar collection device is its placement with regards to the sun. The collector should face the sun which can be tricky because the sun moves across the sky in a arc.

There are many factors dictating the size of the solar collector: the angle of the sun, the presence of clouds, smog, and the length of daylight, the number of people in the household, and the uses of hot water. It is suggested that you find a household similar to yours that uses a solar water heater and get your ideas from them. Consult a solar hot water vendor. You can also purchase solar panels arranged in series or in parallel and start with the minimum you think you might need, then add on later as your particular demands become clear.

Read more in:

Appendix Two—Direct Solar Energy Considerations
Appendix Nine—Light
Appendix Ten—Paint Color
Appendix Thirteen—Solar Water Heaters

Part Four

Appliances

Chapter Eleven

Appliances

Appliances don't consume as much energy as transportation or heating and cooling, but there are things we can do to cut down on the lifetime consumption of 134,342,220,298 calories. If you look at the following table, you can see that the appliances which are turned on most of the day are the ones that gobble up relatively huge quantities of energy. With respect to the desktop computer and the TV, turn them off when not in use, which is the majority of the time in the average household.

The freezer and refrigerator are harder to deal with and require new habits on your part. The easiest ways to conserve energy conservation here is to eat fresh, buy only what you need, and purchase those appliances to fit your needs like a glove.

Table of Appliance Energy Consumption

Appliance	Calories per year
Computer (Desktop)	835,769,573
Refrigerator	445,743,772
Small Freezer	334,307,829
TV—25"	278,589,858
Electric Blanket	99,054,172
Microwave (Large)	92,863,286
100 watt incandescent	92,863,286
Dishwasher	69,647,464
Microwave (Compact)	55,717,972

Appliance	Calories per year
Oven	51,590,714
Hair Dryer	38,693,036
Toaster	28,632,846
Printer	18,572,657
Washing Machine	15,477,214
Coffee Maker	12,381,771
Frying Pan	9,286,329
Hot Plate	9,286,329
Vacuum Cleaner	4,539,983
Waffle Iron	3,714,531
Sink Garbage Disposal	3,482,373
Coffee Pot	3,095,443

The above table isn't complete by any means, and the list keeps growing every day with no regards to their efficiency, or affectivity like the straight razors that require a battery.

The ultimate solution is two pronged:

1- A detailed analysis or needs for electrical or battery operated gadgets that are reported to save time and effort such as remote controls that have an efficiency rating from .008% (non-rechargeable batteries) to 3% (rechargeable batteries)

2- For those that we must have, like a frying pan, we can accommodate them by alternative sources of energy like solar and wind.

Solar energy efficiency. It takes 400 million calories to build a 0.18 kW solar energy power plant, and over ten years it generates in the neighborhood of a trillion calories of energy, giving it an efficiency rating of 1,800,000%.

Wind energy efficiency. It takes nine million calories to build a 10 kW windmill power plant, and over the life of ten years it generates about nine billion calories of power, giving it an efficiency rating of 100,000%.

There are limitations to both these high efficiency sources of energy namely that they are geographically limited to areas with lots of sunlight or wind. They also require a large footprint to implement. For example, to supply all the energy needs of an average home a solar panel larger than the total surface area of the roof is required.

So if you live in an appropriate area and are willing to do without all the modern day gadgets, you can supply all your energy needs using solar, wind or both.

Why haven't solar and wind energy sources been able to be developed in large scale?

Unfortunately, the short business cycles, depreciation schedules, foot print and the cost of borrowing against future earnings makes the prospect of developing solar and wind energy as efficient sources will only happen when the cost of oil, coal and nuclear power becomes prohibitive (50 to 100 years from now, depending on your crystal ball.)

But that depressing prospect shouldn't discourage those of us who look at efficient solutions to our energy needs long term.

Read more in:

Appendix Two—Direct Solar Energy Considerations
Appendix Three—Wind Generated Electric Power Considerations
Appendix Four—Battery Considerations
Appendix Seven—Appliance Wattage Considerations
Appendix Fourteen—Energy Star
Appendix Fifteen—Conversion Table

Part Five

Appendixes

The Numerical Perspective

The power consumption of the human body is on average 1,300 to 2,800 calories per day and may slow to 46 calories per hour while sleeping, and 200 calories per hour for someone doing strenuous physical work.

In the backs of the minds of scientists, accountants, and researchers is an appreciation of the exponential nature of numbers that most of us seldom realize because of the linier form they take in print. For example 46 calories is really

4

6 calories and 146 is really (see next page)

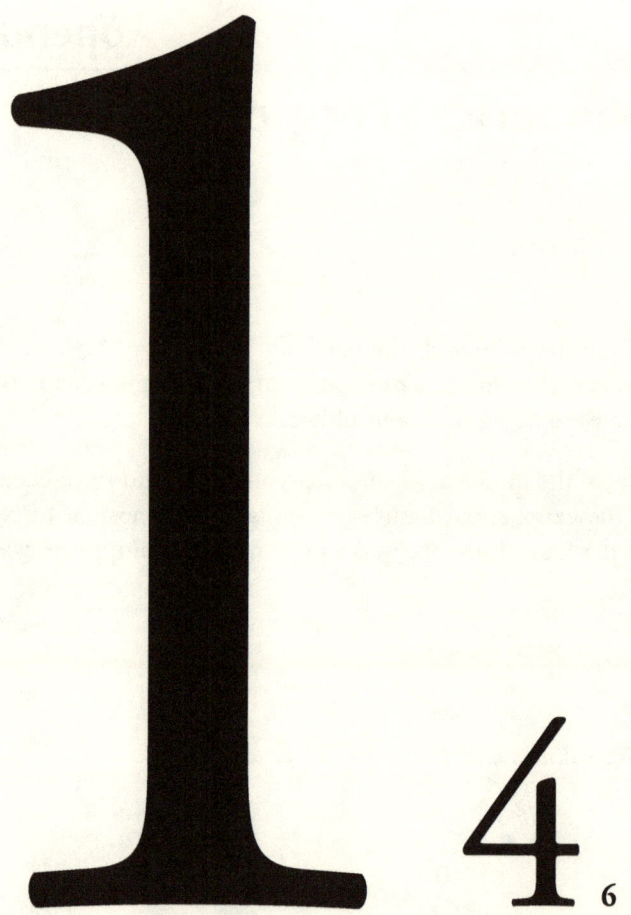

14₆

Large numbers have been used throughout this book and so for a realistic perspective, I will list the sizes of some here:

The number	Its size
800	
8,000	As high as a two story building
80,000	
800,000	Three times taller than Mt. Everest
8,000,000	
80,000,000	
800,000,000	
8,000,000,000	Distance to the moon
80,000,000,000	
800,000,000,000	
8,000,000,000,000	Distance to the Sun

Direct Solar Energy Considerations

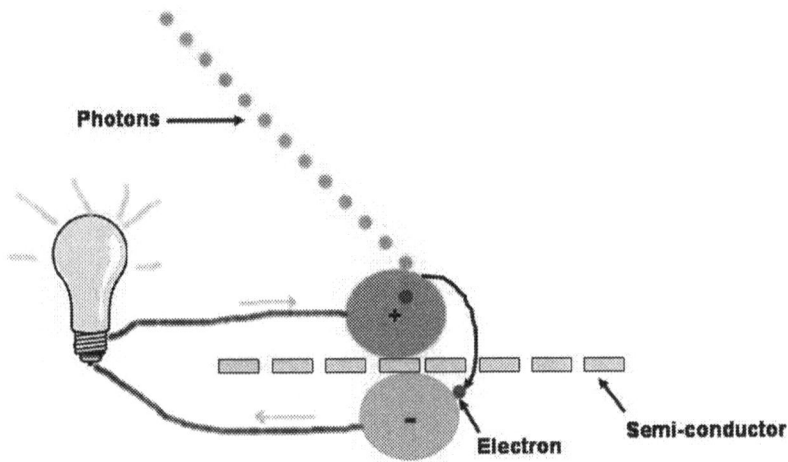

Photovoltaic (PV) devices use light to generate an electrical current. Some materials exhibit a property known as the "photoelectric effect" in which they absorb photons of light and release electrons. When these electrons are captured, an electric current is generated and can be used to do work.

A photovoltaic cell, also called a solar cell, is made of a semiconductor type material, such as silicon. A thin semiconductor wafer is specially treated to create a positive (electron donator) and negative (electron receptor) on each of its sides and when exposed to sunlight, electrons are knocked loose from the atoms in the donator side and travel to the receptor side. If electrical conductors are attached to the donator and receptor sides the electrons create an electrical current as they move back. This current can be used to do work such as power a light bulb or charge a battery.

PV cells have an optimum efficiency of around 5–15% in converting light energy to electrical energy. This is because only a narrow band of the light spectrum (specific wave lengths) can be used to excite electrons out of place in the semiconductor, and different semiconductor materials are excited by different wave lengths.

Environmental factors can decrease this efficiency, such as the movement of the sun, pollution and cloud cover. To overcome the movement of the sun from a vertical position relative to the collecting cell, tracking mounts can be installed to keep the PV cells aimed directly at the sun. Other methods of optimizing sunlight capture include concentrators which gather light from a wider area and reflect it onto the PV cells. In addition, the surface coating of a cell is important in decreasing the amount of photons that bounce off of if thus decreasing its ability to perform.

PV cell systems can stand alone or be connected to an electrical grid run by the electric utility company. They operate only during the day. They have few moving parts, so mechanical maintenance is minimal. They are delicate, however, and require constant cleaning and protection from abrasives (such as dust and sand storms.) Etching from pollution can reduce PV cell life expectance and efficiency.

Basic costs, beginning with a PV system that produces 600 watts and operates several lights, a TV, stereo, microwave and water pump (not at the same time) costs about $8,000.

Table showing cost-benefit structure of a PVC system for generating electricity

Cost

Unit cost	$8,000	
Rating	0.600	kW
Transportation	$300	
Installation	$500	
Life expectancy	10	years
Maintenance	$100	/year
Accessories		
Batteries etc.	$4,000	
Transportation	$300	
Installation	$1,000	
Life Expectance	5	years
Maintenance	$100	/year
Total cost	$15,100	
or	73,249,408	calories

Benefits

Generation	0.60	Peak kW
515,907	Peak Calories	
1,355,803,974,432	Calories/lifetime	

Efficiency	**1,850,942%**

Stand alone systems have battery storage systems and typically cost more. Individual batteries cost about $250 for a heavy duty, 12 volt, 220 Amp hour, deep-cycle battery. (Deep cycle means that the battery can be drained down substantially and frequently.) Larger capacity batteries with higher amp hour ratings, cost more.

Wind Generated Electric Power Considerations

Wind systems can be stand alone or be connected to an electrical grid run by the electric utility company. They are available to generate electricity 24 hours per day depending on geographic wind patterns. They are mechanical, requiring periodic preventative maintenance to improve their efficiency and life expectancy. Increasing size and scale of the generating system can reduce costs and improve efficiency substantially.

A wind power generator consists of a propeller turning an electrical generator, mounted high off the ground. There are two basic ways to capture wind energy, both having to do with the shape of the propeller blades. Broad blades are used to generate low speed torque, and are well-know from the windmills in Holland or the old water pumps used on farms. Electric generating propellers are much slimmer and designed for higher speeds that are required by generators.

Electric generators have an optimum turning speeds at which it generates electricity efficiently. Smart generators stop turning at less than optimum speeds to save wear and tear on mechanical parts. Other smart generators have variable-width propeller blades and transmissions to maintain optimum generator speeds within a wider range of wind speeds. The theoretical maximum efficiency rate for a wind system is 51%.

The total installation cost can be expressed as a function of the wind system's rated electrical capacity. A grid-connected residential-scale system (1–10 kW) generally costs between $2,400 and $3,000 per installed kilowatt. That's $24,000–$30,000 for a 10 kW system.

A moderate-scale (10–100 kW) is more cost-effective, costing between $1,500 and $2,500 per kilowatt. Large-scale systems greater than 100 kW cost in the range of $1,000 to $2,000 per kilowatt, with the lowest costs achieved when multiple units are installed at one location. In general, cost rates decrease as system capacity increases.

Operating costs include maintenance and service, insurance, and applicable taxes (or tax credits.) A rule of thumb estimate for annual operating expenses is 2% to 3% of the initial system cost $.01 to $.02 per kWh of output.

Table showing cost-benefit structure of a wind system for generating electricity

Unit cost	$30,000	
Rating	10	kW
Transportation	$200	
Installation	$300	
Life expectancy	10	years
Maintenance	$100	/year
Accessories		
Batteries etc	$4,000	
Transportation	$300	
Installation	$1,000	
Life Expectance	5	years
Maintenance	$100	/year
Total cost	$37,800	
or	9,028,375	calories
Benefits		
Power generation	10	kW
Power generation	8,598,452	calories
Power generation	9,415,305,378	calories/lifetime
Efficiency	104,286%	

Battery Considerations

Electricity is the flow of electrons through a conductive path. This path is called a circuit. Many dissimilar metallic elements will generate a current through a closed circuit. An example is the earphone radio that operates by connecting two wire leads to different metallic objects like lead and copper plumbing pipes.

Batteries have three parts, a negative anode, a positive cathode, and an electrolyte between them. The cathode and anode (the positive and negative nodes at either end of a traditional battery) are hooked up to an electrical circuit generating an electrical current to do work.

The electrolyte causes a chemical reaction in the battery, and electrons build up in large numbers, causing an imbalance between the anode and the cathode. The electrons want to rearrange themselves to get rid of this instability or difference in charge. By connecting the anode and cathode through a closed circuit, the electrons migrate to the cathode, creating a current that can do work.

These electrochemical processes change the chemical composition in the anode and cathode rendering them less capable of performance, and the battery's efficiency gradually declines.

When you recharge a rechargeable battery, you reverse the directional flow of electrons using another power source, such as solar panels. The electrochemical processes happen in reverse, and the anode and cathode are restored to their original state and can again provide full power. The electrolyte, anode and cathode parts of a rechargeable battery are made of different materials than a non-rechargeable battery.

Rechargeable batteries have an efficiency of around 75% to 85%. This efficiency is measured by dividing the output power by the power used to recharge it, and generally don't reflect the cost of the battery itself. When the initial cost of a battery and its disposal are considered, a battery's efficiency declines rapidly as can be seen in the following two tables that demonstrate the efficiencies of batteries purc.hased from a local grocery store.

Non-Rechargeable Batteries Costs and Efficiency

	Cost	Calories	kW
Purchase cost/pound	$15.008	72,803.120	
Disposal Cost/pound	$5.660	27,456.400	
Power generated/pound		7.643	0.032
Efficiency	0.008%		

Rechargeable Batteries Costs and Efficiency

	Cost	Calories	kW
Purchase cost/pound	$33.576	162,875.636	
Disposal Cost/pound	$5.660	27,456.400	
Power generated/pound		22.929	0.096
Power to Recharge		45.858	
Number of Charges		300.000	
Total Power Input		13,757.524	
Total Power Output		6,878.762	
Efficiency	3.370%		

Batteries reach peek performance in power delivery when they are fully charged. This performance declines as it discharges. Heavy duty, or deep cycle batteries maintain high performance even when they are half or a quarter of the way charged. Deep cycle batteries are used when the application can only charge them up part of the time, such as with solar power cells.

Batteries need to have a minimum charge, also known as a "seed charge," in order to be recharged effectively. Batteries also have a relative short life span of about five years due to their chemical composition, and proximity of the elements that constitute it. These changes accelerate when the battery is less than fully charged.

Due to the chemical nature of batteries, they should to be disposed of properly or recycled when possible when they stop being useful, to avoid releasing the many toxic and corrosive compounds they contain. This adds to their overall cost and the cost of the applications they comprise.

Fuel Cell Considerations

A fuel cell combines two elements to produce an electrical current. One fuel cell having much potential is the oxygen-hydrogen fuel cell developed by the NASA. The advantage of the fuel cell in space is that it not only produces electricity for the spacecraft, but it also produces water as a secondary end product reducing the need to carry a full load of needed drinking water on space missions.

This secondary byproduct makes the fuel cell attractive to researchers who are looking for ways to reduce pollution in autos, and the prospect of a car that produces water vapor as exhaust is still tantalizing.

Two steps are required to generate electricity with a fuel cell. The first is the separation of water into its two elements (oxygen and hydrogen.) The second is the re-combination of oxygen and hydrogen to generate electricity using a catalyst. The first step requires energy as an input while the second generates energy. Unfortunately, this process as a whole consumes more energy than it produces compared to a battery, and today the energy needed to separate water into its elements comes from fuels that cause air pollution, thus defeating the whole purpose of the fuel cell as a replacement source of power for the auto.

This is true unless another non-polluting source of energy is found to separate water into its elements. One such source is the Photovoltaic cell that converts solar energy into electrical energy. If and when that happens, the fuel cell will come in direct competition with the battery in powering autos.

The pros of the fuel cell:

1- Oxygen is portable
2- Dispensing of oxygen is much like propane

Cons of the fuel cell:

1-   Potentially hazardous (hydrogen is explosive)
2-   Less efficient than batteries

The pros of the battery:

1-   The net efficiency is greater than the fuel cell
2-   The technology is available
3-   Safer

Cons of batteries:

1-   They are bulky
2-   Require long periods to recharge
3-   Disposal is hazardous and pollutes

R-Value Considerations

The R-value in grading an insulation performance comes from work in electrical conductance. In that field, three factors influence the conductance of electricity: surface area of the conductor, difference in energy levels, and its temperature.

The equation that describes this is:

$$U = W/m^2 * K$$

Where
* is the multiplication sign
U is the conductance
W is energy measured in Watts
m^2 is the area of the surface of a wire for example in metric
and K is the temperature in Kelvin.

Since the inverse of conductance is resistance, the folks in the insulation business inverted the equation and so it looks like this:

$$R = m^2 * K/W \text{ (where R is resistance to energy transfer)}$$

The equation and was modified and K was replace with the difference in temperature between the two sides through which energy is transferred. For example, if the interior of your home has a temperature of 20°C and the outside is at 10°C the temperature difference is 10°C. Then, if the walls are insulated to R = 2, 5 Watts of energy will be lost every second for every square meter of wall.
The equation for the above problem is:-

$$W = m^2 * K/R$$

A non-metric equation replaced W with BTU (British thermal units), m^2 with feet^2, and the temperature is measured in Fahrenheit degrees. The equation looks like this:

R = ft^2 * ^{o}F h/Btu

The relationship between the metric and non-metric R values is as follows:

R-non-metric = approximately 0.1761 R-metric

Further, the non-metric R-value is expressed in square inches.

Table of R values for various substances in alphabetical order

Material	Metric R-Value	Non-metric R-Value
Air (300 K, 100 kPa)	38.168	6.721
Aluminum	0.005	0.001
Brass	0.009	0.002
Copper	0.003	0.000
Diamond	0.001	0.000
Fiberglass	17.036	3.000
Fiberglass	25.000	4.403
Glass	0.930	0.164
Gold	0.003	0.001
Ice	0.526	0.093
Iron	0.012	0.002
Isocyanurate foam	47.132	8.300
Lead	0.029	0.005
Loose cellulose	17.036	3.000
Mercury	0.119	0.021
Phenolic foam	39.750	7.000
Platinum	0.014	0.002
Polystyrene		
Expanded bead board	33.333	5.870
Quartz (273K)	0.106	0.019

Material	Metric R-Value	Non-metric R-Value
Rock wool	17.036	3.000
Silica aero-gel	58.824	10.359
Silver	0.002	0.000
Snow	5.679	1.000
Steel	0.020	0.004
Straw bales	8.234	1.450
Styrofoam	46.512	8.191
Water	1.650	0.290
Wood		
(Average balsa/fir)	5.128	0.903
Wool	22.222	3.913

Appliance Wattage Considerations

The formal definition of a Watt (W) is: A unit of power equal to 1 joule (0.2388459 calories) per second; the power dissipated by a current of 1 ampere flowing across a resistance of 1 ohm.

The makers of appliances wanted to standardize their ratings to the KiloWattHour (KWh) nomenclature used by power companies and utilities, because most appliances sip electricity at very low rates (Watts) per second. So they invented a rating that describes the power (Watts) consumption per hour instead of per second. And so an appliance rating of 20 W is really a 20 Watthour consuming appliance, to fit it into an equation from a power utility, simply divide by 1000.

Appliance Watt Usage Conversion Table

Wattage	KWh	Ounces of Oil	Ounces of Coal	Calories
10	0.010	0.010	0.130	2
15	0.015	0.015	0.195	4
20	0.020	0.020	0.260	5
25	0.025	0.025	0.325	6
30	0.030	0.030	0.390	7
35	0.035	0.035	0.455	8
40	0.040	0.040	0.520	10
45	0.045	0.045	0.585	11

Wattage	KWh	Ounces of Oil	Ounces of Coal	Calories
50	0.050	0.050	0.650	12
55	0.055	0.055	0.715	13
60	0.060	0.060	0.780	14
65	0.065	0.065	0.845	16
70	0.070	0.070	0.910	17
75	0.075	0.075	0.975	18
80	0.080	0.080	1.040	19
85	0.085	0.085	1.105	20
90	0.090	0.090	1.170	21
95	0.095	0.095	1.235	23
100	0.100	0.100	1.300	24
200	0.200	0.200	2.600	48
300	0.300	0.300	3.900	72
400	0.400	0.400	5.200	96
500	0.500	0.500	6.500	119
600	0.600	0.600	7.800	143
700	0.700	0.700	9.100	167
800	0.800	0.800	10.400	191
900	0.900	0.900	11.700	215
1000	1.000	1.000	13.000	239
2000	2.000	2.000	26.000	478
3000	3.000	3.000	39.000	717
4000	4.000	4.000	52.000	955
5000	5.000	5.000	65.000	1,194
6000	6.000	6.000	78.000	1,433
7000	7.000	7.000	91.000	1,672
8000	8.000	8.000	104.000	1,911
9000	9.000	9.000	117.000	2,150
10000	10.000	10.000	130.000	2,388

Planting Distances for Trees to Shade a House (in feet)

Height of tree =

      Step 1 = 90 **plus** angle of sun from horizon

      Step 2 = Distance from house **times** angle of sun from horizon

      Step 3 = Step 2 **divided by** Step 1

      Step 4 = Number of floors **times** 9

      Step 5 = Step 3 **plus** Step 4

Distance from house =

      90

Minus

      Angle of sun from the horizon

      ## times

      Tree height

      ## divide by

      Angle of sun from horizon

      ## plus

      One half the girth of the foliage

Light

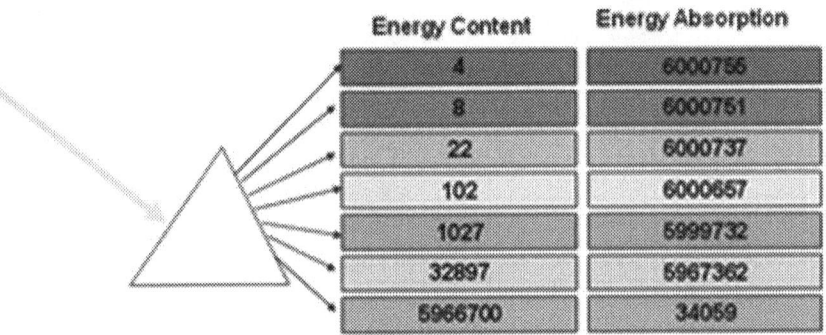

Energy Content	Energy Absorption
4	6000756
8	6000751
22	6000737
102	6000657
1027	5999732
32897	5967362
5966700	34059

Visible light is part of the electromagnetic spectrum of energetic rays that exists in the universe. Most of this energy never gets past the three mile point above the Earth's surface. Almost the only energy that makes it to its surface is visible light that we can detect with our eyes.

We can use a prism to fracture sunlight into its colorful components. The fracturing process illustrates the energy content of the different colors. Light of the red color has an amount of energy associated with it and so does the color blue. The color spectrum can be divided into seven or so colors. They are: red, orange, yellow, yellow-green, green, green-blue, and blue.

If six million calories of light were separated into their components according to each color's energy level, the distribution of colors would be broken down in the following table. Note that the energy value of each color going down the "Energy

Content" column of colors is exponential (meaning that the numbers get bigger by factor of the previous color.)

Color	Energy Content	Energy Absorbtion
Red	4	6000755
Orange	8	6000751
Yellow	22	6000737
Yellow-Green	102	6000657
Green	1027	5999732
Green-Blue	32897	5967362
Blue	5966700	34059

If you see the red, your eyes are receiving energy of about 4 calories. If you see yellow your eyes are receiving 22 calories, and blue gives your eyes about 5.9 million calories.

If the object that you are looking at appears red, and is exposed to the sun's direct light, you might ask what happened to the rest of 6 million calories of energy. The answer is most likely that they were absorbed by the object, as in the next diagram and quantified by the column in the table above.

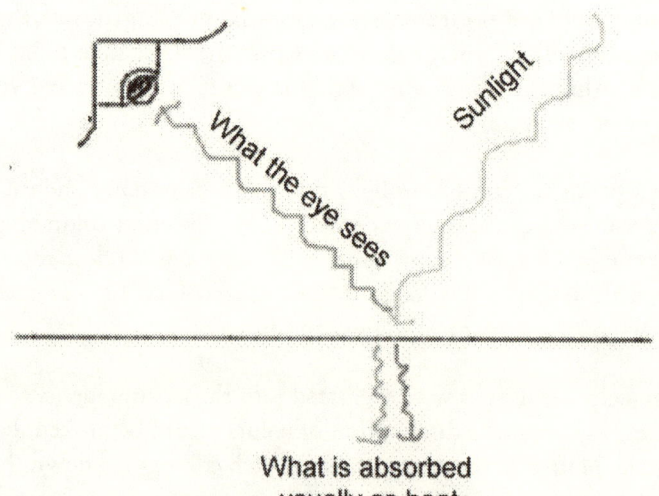

What is absorbed
usually as heat

If a building is painted red, it absorbs most of the six million calories that are directed to it and it will get hot in a hurry. If the building is painted blue, only a fraction of the energy is absorbed and the building will be a lot cooler.

So far we have been discussing an imaginary world. In the real world paint starts out as white and the color pigment is mixed in. The ratio of pigment to paint is about five parts per hundred. So you can divide the absorption numbers in the diagram by one or two hundred to get a more realistic estimate.

Did I tell you that sunlight is made up of all colors together, and it is for all intent and purposes white?

Paint Color

Paint doesn't follow the color pattern in the previous appendix. Paints are not all true or pure pigments. They usually consist of a base substance, a brightner or darkner, and a pigment or a combination of pigments. These all have an effect on how much light is absorbed or reflected back to the eye.

When considering what color to chose, it helps to spend some time experimenting. Dark paint absorbs a lot of energy that is transformed into heat. The following is an experiment to measure energy and heat absorbtion.

Imagine a cardboard box with one side repeatedly painted with different colors and exposed to a halogen light bulb to simulate exposure to the sun for about an hour. The box is about a foot wide, half a foot high and a half a foot deep. The box sits a foot away from the light source. A thermometer sits inside to measure temperatures.

The table bellow shows the results of the heat absorbtion experiment and translates the measured results in temperature rise to that of the temperature rise in a 1,500 square foot home with cardboard walls.

Heat Absorbtion in degrees

Color	Experimental Results	Expected Rise in Temperature in a 1500 Sq Ft House
White	4	10
Black	15	120
Navy Blue	15	120
Dark Green	12	90
Red	13	100
Yellow	10	70

Of course we don't live in cardboard houses and so the temperature would never increase to the extent that is shown in the table. The R value that our external walls are constructed to must be considered.

To extrapolate your results to the finished product, you can use these formulas:

Coefficient (Cx) = (time exposed **times**
width **times**
height **times**
temperature change **divide by**
volume) of box

Change of temperature in house = Cx **times**
(volume **divide by**
(exposure time **times**
width **times**
height)) of house

Appendix Eleven

EPA Fuel Economy Auto Testing

Each vehicle in the EPA fuel economy laboratory is placed in a room with large rollers on the floor. A simulation program drives the vehicle along a regiment of starts, stops, speeding ups, and slowing downs. All vehicles are run through the same "city" and "highway" simulations.

These are highly artificial circumstances (for example, they don't measure wind resistance) and therefore the agency adjust the results for more realistic conditions in the year 2008.

Fuel economy ratings shouldn't be taken at face value, but rather used to compare different vehicles only in terms of effects of weight, drag and engine performance. If you are looking for a vehicle to purchase, you find two that you like and one is rated at 21 and the other at 31 miles per gallon, chances are that the latter is more fuel efficient. Actual mileage will depend on your style of driving.

Heat Exchangers

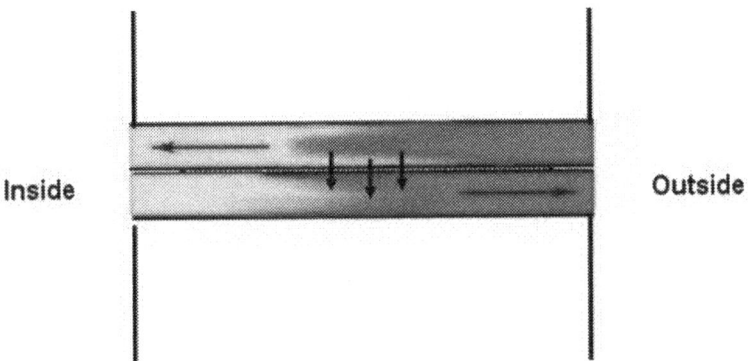

Inside   Outside

A heat exchanger is used to conserve heat when ventilating a house. The principle of a heat exchanger was found in husky dogs in Alaska by biologists. They discovered that arteries in the legs of huskies run close and opposite to the veins, to conserve heat around their vital organs.

A simple heat exchanger consists of two pipes directing air or water in opposite directions. The neat thing about them is that the temperatures of air or water are pretty close at both ends if the pipes are long enough. The length of the pipes allows the hot and cold medium to be coupled long enough to allow the heat to be transferred from one to the other.

There are many heat exchange designs depending on the application to which they are to be applied. You can purchase heat exchangers that can conserve anywhere from 50% to 90% of the heat.

Heat exchangers can be used to keep buildings cool as well as hot using the same principle.

Appendix Thirteen

Solar Water Heaters

Solar water heaters consist of a solar panel used to collect heat from the sun's rays, and a water storage tank and are made of non corrosive material like brass or stainless steel. You can purchase a passive heater, meaning that it has no moving parts and depends on physics to circulate hot water from the panels to the tank.

The critical part of the solar water heater is the collection panel. It is usually a coil or array of tubes painted black. The most efficient are enclosed in a window-like box with double or triple glazed glass. The coil or tubes are supported by a metallic plate that is also painted black.

A passive water heater has a water tank above the collector panel and depends on hot water rising from the top of the panel to the tank and being replaced by cooler water from the bottom. Other solar water heaters use a water pump to circulate the water from the tank to the collector panels allowing the placement of each of the tank or the collector panels in any position.

As you go north, the ability to raise the water temperatures to desirable levels diminishes because of colder weather and cloud cover. In this event, the solar water heater is used to warm up the water before it goes into the electric or gas hot water heater, reducing the cost of hot water.

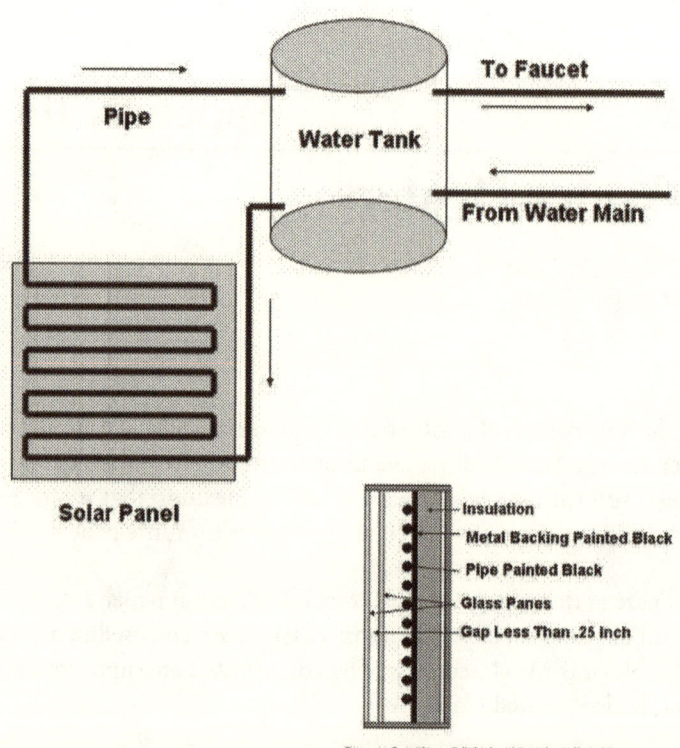

Pipe

To Faucet

Water Tank

From Water Main

Solar Panel

Insulation

Metal Backing Painted Black

Pipe Painted Black

Glass Panes

Gap Less Than .25 Inch

Cross Section Of Solar Heating Panel

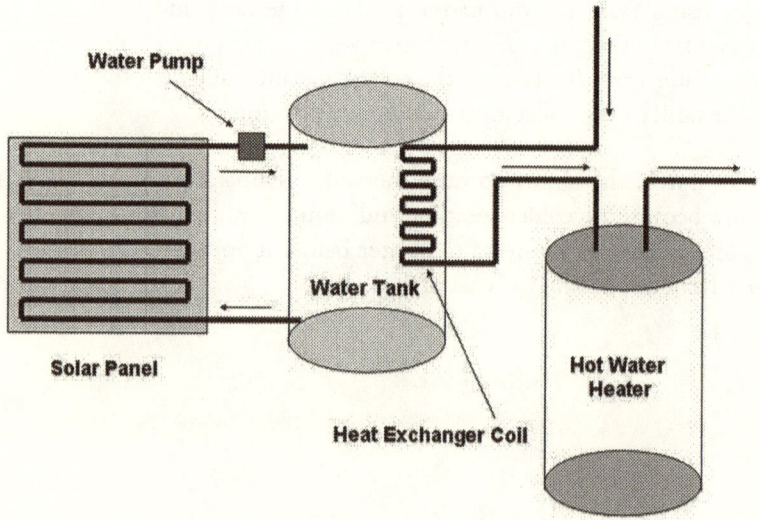

Water Pump

Water Tank

Hot Water Heater

Solar Panel

Heat Exchanger Coil

Energy Star

ENERGY STAR is a dynamic government/industry partnership that offers businesses and consumers energy-efficient solutions, making it easy to save money while protecting the environment for future generations.

In 1992, the US Environmental Protection Agency (EPA) introduced ENERGY STAR as a voluntary labeling program designed to identify and promote energy-efficient products to reduce greenhouse gas emissions. Computers and monitors were the first products labeled. Through 1995, EPA expanded the label to additional office equipment products and residential heating and cooling equipment. In 1996, EPA partnered with the US Department of Energy for particular product categories. The ENERGY STAR label is now on major appliances, office equipment, lighting, home electronics and more. EPA has also extended the label to cover new homes and commercial and industrial buildings.

Conversion Table

Unit	Conversion	Unit
1 barrel =	158987.3	cc
1 barrel of oil =	1410579000	calories
1 Btu =	251.9968	calorie
1 Btu =	3413	Kilowatt-hours
1 cm =	0.393700787	inch
1 cubic foot gas =	258296.72	calories
1 Degree Centegrade	(°F-32)/1.78	Degree Fehrenheight
1 Degree Fehrenheight	32 + (°C * 1.78)	Degree Centegrade
1 dollar =	23509650	calories
1 foot =	0.3048	meter
1 foot =	30.48	cm
1 Foot =	12	inches
1 foot^3	1728	inch^3
1 Foot^3 air	0.07608	Pounds at sea level at 59Degrees
1 Gallon =	3.785	liters
1 Gallon =	3.5817455	liquid quarts
1 Gallon =	4	quarts
1 Gallon	231	inch^3
1 Gallon	7.480519481	feet^3
1 Gram =	0.035273369	Ounces
1 Inch =	2.54	cm
1 joule =	0.2388459	calorie

Unit	Conversion	Unit
1 kilo meter =	0.621349571	mile
1 Kilogram =	2.204585538	pounds
1 KW	238.8459	calories per second
1 KWh	3412.127614	Btu
1 KWh	859845.24	calories
1 Liter	0.264200793	gallons
1 meter	3.280839895	feet
1 meter =	3.280839895	feet
1 meter =	39.37007874	inch
1 meter =	1.093613298	yard
1 Meter^2	1550.0031	inch^2
1 Meter^3 air	1.218705193	Kilograms at 59 degrees
1 mile	5280	feet
1 Mile =	1.6094	km
1 M ton =	1000	kg
1 Ounce =	28.35	grams
1 Pound =	0.4536	kg
1 quad =	2.51997E+17	calories
1 Quart =	2	pints
1 Quart Dry =	1.101	liters
1 Quart Liquid =	0.9463	liters
1 cubic foot =	0.028316847	cubic meters
1 cubic meter =	35.31466672	cubic feet
1 R (non SI foot)	0.1761	R (SI = m2 * K/W)
1 R (non SI inch)	304.3008	R (SI = m2 * K/W)
1 short ton =	907.1847	kg
1 Sq meter	10.76391042	Sq Ft
1 Therm =	100000	Btu
1 ton =	907.1847	kg
1 ton coal =	25000000	Btu
1 ton coal =	6299920000	calories
1 Ton English =	2240	pounds
1 Ton Metric =	2204.6	pounds
1 Ton US =	2000	pounds

Unit	Conversion	Unit
1 watt	1	Joule/second
1 watt	0.2388459	calories per second
1 Yard =	0.9144	meter
1 Yard =	3	feet

978-0-595-39263-6
0-595-39263-6

www.ingramcontent.com/pod-product-compliance
Lightning Source LLC
Chambersburg PA
CBHW021545200526
45163CB00015B/1743